# Zoom for Teachers (2020 and Beyond)

A Beginner to Expert User Guide to Master the Use of Zoom for Meetings, Virtual Classes, Businesses, Video Conferencing, and Webinars

By

Nicholas Scott

## Disclaimer

This publication is designed to provide competent and reliable information regarding the subject matter covered. However, the views expressed in this publication are those of the author, and should not be taken as expert instruction or official advice from Google. The reader is responsible for his or her own actions.

The author hereby disclaims any responsibility or liability whatsoever that is incurred from the use or application of the contents of this publication by the purchaser or reader.

1

# Books By The Same Author

*Apple Watch 5 Manual (2020 Edition)*

*How To Cancel Kindle Unlimited*

*How to Delete Books From Your Kindle Library*

*Google Classroom 2020 and Beyond*

# About The Author

Nicholas Scott is a tech geek fanatic, with over 15 years of experience working in the tech space, right in the heart of Silicon Valley. Nick, as friends fondly call him, holds a Bachelor's degree in Software Engineering from MIT and a Master's degree in Information Technology from Stanford.

His hobby and passion revolves around geeking about technologies, and writing tech and computer guides for just about any trends in information technology.

He is married to a lovely wife and has two kids.

# Table of Contents

# Introduction

A few years ago, not many people would have believed the world would come as far as it has today. All thanks to technology, many things have changed and evolved for the better.

One of the major areas where the effect and impact of the advancements in technology are felt so much is in online meetings and the possibilities that have been unleashed through the power of the internet.

Decades ago, communication was stilted in the sense that there had to be a lot of time involved for communication to take place. One person had to write the other from one end of the world and hoped that his letter is received on time. He also hoped that he could get an answer from the person at the other end of the communication chain. For meetings to occur, people had to travel from different parts of the world to be congregated at a place; where they got to interact with themselves and do the things for which they came together for.

This was a major challenge because sometimes, it was just not possible to get some of the people that needed to be in the meeting to be present physically. At other times, getting them to be physically present was not just worth the hassles involved. Because of this, many people had to pass on many opportunities, which took its toll on communication and the way the world worked.

Good thing technology stepped in to save the day!

A few years later, many tools that enabled teleconferencing and distance social relations started hitting the consumer market. For example, Bell Labs created the first phone that was capable of transmitting a video signal across the U.S in 1964. This phone was the result of a ton of work, and although it was universally praised for the work it was able to do (in terms of the picture quality and call interface), there was a major pitfall with it, and many more of the devices and tools that were innovated at the time for teleconferencing and long-distance social interactions.

The tool was very expensive. It could only be afforded by the elite, and sometimes, it took a level of wealth and influence to have access to these tools. So while this was a great achievement in science and technology, it was still limited.

Thankfully, technology was still in the business of saving the day.

Fast track to today's world, there are a ton of tools and softwares that have been developed with the sole intent of bridging this major communication divide. In the space of milliseconds, a person can, from one end of the world, connect with another person leagues away from where he is. Technology, and the power of the internet, have made it possible that communication and real-time distance interaction is almost as seamless/effortless as man's breathing. It does not have to take forever for information to be passed from one person to another, despite how far away they are from themselves. Communication is now affordable with a video conferencing tool, and people do not have to pay through their nostrils to interact with themselves.

All it takes now is a few clicks on a device connected to the internet, and the rest is history. Thanks to technology for the innovation of these tools that have made the world an easier place.

This book is going to be centered around one of them; the Zoom app.

In a nutshell, Zoom is a teleconferencing tool that allows its users the liberty to communicate in real-time over the internet, using a peer-to-peer, cloud-based platform. Leveraging this tool allows you to choose whether you want to meet with people via video or audio. The Zoom app allows people to communicate across distances without leaving their homes' comfort or wherever they find themselves.

From school meetings to lectures, corporate events, distance education, and social relations, Zoom offers its users the flexibility to interact seamlessly without going through many hassles.

This book takes an in-depth look at the Zoom tool, especially how it relates to teaching effectiveness. As a teacher, you need to constantly communicate with your students, notwithstanding how close to you they are per time. This book covers the basics of what you need to know about the Zoom application and by the time you are done with this guide, you will discover;

- How to master the Zoom tool to make the most of it for your meetings and virtual events.
- How to make use of Zoom for video conferencing and for hosting webinars.
- The different pricing structures that you can subscribe to (although we have said that the tool is not as costly as videoconferencing tools before it, it is not entirely free either).
- The things you need to consider before selecting a Zoom plan and what you stand to gain from each of the plans.
- How to set up and make proper use of the Zoom app to ensure your virtual meetings (classes for teachers and students) are as successful as they can be.

There are still a lot more things your Zoom app can do for you that you probably did not know of. Well, you are about to find out all those cool tips and tricks and more stuff that Zoom brings to the table. So, read on because you are about to Zoom off (pun intended).

# Chapter 1

## A -Z of Zoom

### What is Zoom?

Zoom is a tool that provides teleconferencing, telecommuting, distance education, and social learning through video telephony and online chat services delivered over a cloud-based peer-to-peer communication system.

Simply put, Zoom is a web-based video conferencing tool that allows its users to meet online either through audio means or video means. A host of devices supports this tool as it has mobile apps for different platforms, desktop versions and a web-based version of the tool. Across its millions of users scattered across the world, this tool is highly famed and celebrated because of its easy-to-use interface, high-quality HD video and sound support. The tools and features embedded in it also allow for easy collaboration.

As a teacher or a tutor of any kind, chances are that you usually have a need to meet with people or even host training for a large number of people in different locations. If you have ever found yourself in this position or find yourself in this position often, Zoom is one tool that can help you interact with all the people you need to, and still be able to communicate with them without having to break into a sweat. This tool enhances connectedness and promotes collaboration and long-distance learning.

**Who Can Use Zoom?**

One of the Zoom app's major pros is that it has a plan/feature for all kinds of users. Zoom Video Communications (the makers and owners of this tool) put a lot into consideration when they produced and launched their tool into the consumer market. As a result of their thinking and strategic planning, Zoom caters to almost all kinds of people.

If you are a

- Teacher
- Business owner
- Online coach
- Online Consultant
- Leader of any kind, including political and religious leader;
- Any kind of person that needs to congregate people to be able to meet and interact with them

14

for any reason at all (including the need to hold a virtual event)

Then the Zoom app is for you.

There are different plans that have been created to suit your different needs. Because the developers of this tool know that not all people who need it will have the same kinds of needs, they were able to come up with different options to suit every person who will have the need to make use of Zoom at some point. The different plans and packages are explored in detail in later parts of this book.

## Main Features of Zoom

There are a few features that make the Zoom app highly acclaimed and loved by millions of people around the world. These are the features responsible for the great experience users have when they get to use Zoom for different purposes.

Here are the features and what they do.

1. Video conferencing; this is one of the features of the application that makes it very useful to the world at large. Zoom is a tool that allows users to communicate with themselves in real-time, using high-quality video

content and graphics. Coupled with this feature is the screen sharing feature that enhances corporate presentations and keynote speeches. In addition to the screen sharing feature, Zoom allows users to select whether they want to share their whole desktop screen (which means that everyone present in the meeting will see whatever the presenter has on the screen of his device), or if he just wants to share screen from one application. This reinforces users' privacy and grants them some more control over what the people present in a meeting have access to.

2. Another main feature of the Zoom app is that sessions can be upgraded and expanded to accommodate a lot more people. When expanded, Zoom rooms can hold up to 300 people, while Zoom webinars can be attended at the same time by up to 10,000 viewers.

3. Recording feature; although virtual meetings are a thing in today's world, there is always that chance that life can get in the way of someone who was supposed to attend a meeting. This application's record feature permits the host of the meeting to record the meeting and save it to a storage device. This way, the meeting can be accessed later, which is handy for reference

purposes, or for those who were unable to attend the meeting to catch up with whatever happened in the meeting at their own convenience. With just a few clicks, recordings can be backed to a cloud storage, and this feature reduces the chances of losing data, even when these files need to be accessed using other devices.

4. Collaboration; the Zoom app is built for collaboration. Features like the chat sections, breakout rooms, and annotation sections allow for collaboration of different kinds.

5. Background changing; this is arguably one of the most creative features of the Zoom application. This feature allows you to replace your backdrop with a selected picture. This way, you can have the courage to share a video without having to stress out about whether or not your background looks good. There are a ton of pictures that can be used as backdrops, and all you need to do is select the one you want to make use of.

6. Beauty mode; this is another creative feature that is found in the Zoom application. Usually, it is not too strange to see picture-taking applications that allow the user to touch up their appearance before taking photos.

However, Zoom was one of the first video conferencing tools to incorporate this feature into its platform. With a few customization options in this feature, the user has the liberty to fix up his appearance and make sure that he is looking his best for whatever meeting he is attending.

7. The gallery view tool allows you the luxury of having a wider view of the meeting. With this, you can have an overview of all the participants of the meeting, which is also great for collaboration.

## Devices Supported By Zoom

The Zoom application has a lot of requirements that form the basis for the successful and seamless operation of this tool. There are a few devices and specifications that must be met to use this application. They are;

## Supported Tablets and Mobile Devices

- All iOS and Android devices.
- Windows Surface PRO 3 running win 10.
- Blackberry devices.

## Supported Operating Systems

The Zoom application will function properly on your PC device if it supports any of these operating systems.

- Mac OS with MacOS 10.6.8 /(Snow Leopard) or later.
- Windows 10
- Windows 7, 8, or 8.1
- Windows XP with SP3 or later versions.
- Windows Vista with SP1 or later.
- Ubuntu 12.04 or higher.
- Mint 17.1 or higher.
- Fedora 21 or higher.
- OpenSUSE 13.2 or higher.
- ArchLinux (64-bit only).
- Red Hat Enterprise Linux 6.4 or higher.

Devices that run with any of these operating systems will be a good fit for the Zoom app. Let this serve as a guide as you pick up your next device. If you want to use the Zoom app on your device, you need to be sure that it runs on any of these operating systems.

**Supported Browsers**

It is necessary to know what browsers the application supports. This is to enable you to prepare better if you are looking to access the web version of Zoom and make use of it for your meetings. Here are a list of the browsers with which you can access the Zoom platform.

- Windows; IE7+, Firefox, Google Chrome, Safari5+.
- On Mac computers; Safari5+, Firefox, Google Chrome.
- On Linux devices; Firefox, Google Chrome.

## Pros and Cons of Zoom

Just like every other software you can possibly think of, the Zoom application comes with its own fair share of pros and cons. Here is a list of some of the major advantages and downsides of the application.

## Zoom Pros

1. Zoom's on-screen whiteboard feature allows you and your team to easily brainstorm ideas. This is a great collaboration feature as it allows teams and organizational units to develop solutions as a body.

2. The live video chat, alongside all its included features, allows for meetings to be held in real-time, and for people from across the world to communicate

seamlessly. Other accompanying features like the chat section and breakout session feature allows for a closer bonding and seamless collaboration.

3. Zoom offers the host of a meeting the opportunity to have a closer look at the back end analytics. This feature allows him to know important data and metrics of the people that attended any meeting, including those that stayed through the meeting, active participants, and other important details.

4. One of the major benefits of using Zoom is that it is relatively affordable. The Zoom app allows users to subscribe to a free plan, which allows them to host meetings with up to 100 people, for up to 15 minutes at a stretch. This is a great feature integrated into the application because even those with a tight budget can profit from the platform, at least to an extent. As time progresses, and as the user begins to have access to more funds and as the needs arise, he can choose to upgrade to paid versions to have access to more features of the application.

5. The Zoom app has features that make it quite durable and a delight for users. For example, the application allows the meeting host to record his meeting and save

the recording to a storage device. This feature alone has proven to be quite handy for various institutions and learning organizations, affording them the ability to streamline their activities. Also, tools like screen-sharing foster professionalism as keynote presentations in teleconferencing sessions are possible, and also screen-sharing is a great way to make sure that many people benefit from the meeting.

6. Some other benefits of using Zoom include access to help and support teams that can walk you through challenges and difficulties, tutorials that guide beginners and even intermediate users to become much better with making use of the software, and access to a community of other Zoom users. This community is great for fostering collaborations and real-time solutions to issues that may arise as users make use of the application.

**Zoom Cons**

1. As your team increases and you begin to move across plans, Zoom becomes a bit on the high side. As you choose to migrate to higher levels like Zoom pro or Zoom enterprise, the cost of accessing and subscribing to the service per month begins to increase. This is because the application is paid for by the host; hence,

the larger the team, the more expensive it gets to use Zoom.

2. Users have complained that there is a level of unpredictability associated with Zoom. This implies that the application can get to a point where the video quality and audio quality begins to deteriorate. In extreme cases, this can affect the steady flow of a meeting and in mild cases, it can cause a glitch in the quality and presentation of the meeting.

3. Although the application integrates a lot of helpful features and tools that are needed for business growth and collaborations, it can sometimes be difficult to navigate these features - especially for users that are not too technologically inclined. Features like the annotations and whiteboard animations can take a lot for the users to wrap their heads around them. The result of this is that people will most likely tend to skip the features they do not know how to use, hence limiting the potentials of the application.

4. Zoom is heavily dependent on the strength and health of internet connectivity. Although the application is web-based, it takes a strong internet connection to attend and have a smooth meeting experience. The result of this is that people who need to join a meeting,

but for some reason, are in places where they have limited internet strength, will most likely experience glitches in accessing the Zoom application to attend meetings.

# Chapter 2

## Zoom Account Types and Plans

### Zoom Paid Users

When you start making use of the Zoom app at first, you are given the option to be a free user or to upgrade to any of the paid versions. Depending on your needs and the size of your team, you can decide to stick to the free plan for the time being, but as the days unfold and as your organization begins to grow and pull in more revenue, there is every chance that you will want to upgrade to a paid plan.

As expected, the paid plans come with a lot of features and perks that are not available to those that use Zoom's free plan. Here are a few of the advantages available to those who use paid versions of the Zoom app.

1. With the paid version, you can be able to host a meeting with more than 100 people in attendance at the same time, and this meeting can last for more than 45 minutes. The free plans allow you to only meet with up to 100 people (and no more) for less than 45 minutes.

Suppose you usually organize classes or teach in sessions that typically have more than 100 people in attendance, and these sessions usually last for more than 45 minutes. In that case, you may want to go for any of the paid packages that meet your needs.

2. With the paid version of the application, you are allowed to record your meetings directly to the cloud. This feature makes sure that you have extra space on your device, and the sharing experience with those that need to have access to the recordings of the meetings becomes easier. The paid version of Zoom gives the user access to up to 1GB of cloud recording space, and the user does not have to pay for this space separately.

3. One of the perks of having a pro (paid) plan on Zoom is that it allows you the liberty to live stream on social media. When you have paid for a pro plan and you have received the pro license, you can tweak the settings and with a few extensions, stream the live meeting to other channels like social media. The result of this is that you can potentially expand your reach and get across to more people with your content.

4. The paid version of Zoom opens you up to a whole world of possibilities. Upgrading to a paid package allows you to have access to extra features including;

- Zoom rooms and Zoom connectors.
- Zoom video webinars that allow you to host up to 10,000 people.
- Better meeting licenses and audio plans.
- Meeting branding and further customization tools.
- Access to Zoom's API; which is a service that allows developers to be able to customize your organization's Zoom experience and build an app for you. You can add many useful extensions and features to the app like activating chatbots, embedding Zoom into your organization's solutions, and much more.

## Zoom Free Users

Zoom free users have access to some of the features that are available in the app. As earlier reiterated, the free plan can be useful to you, depending on your needs and on the size of the audience that needs to be present in your meetings.

Here is a rundown of the features that are available in Zoom's free package;

- The user gets an unlimited number of one-on-one meetings that can last for up to 24 hours at a stretch.
- The user has access to host up to 100 people in a group meeting that can last for up to 40 minutes at a go. After 40 minutes, the meeting will end automatically.
- The free Zoom plan allows the host to record his meeting, but it does not offer the cloud storage options available in the paid plans.
- This plan also affords the user the video and audio conferencing features, but they must be kept within the range of time discussed above.
- The host can share his desktop or application screen.

## Zoom Plans and Pricing

Here are the plans available in the Zoom app, and how much they are worth.

| No. | Plan | Price |
| --- | --- | --- |
| 1 | Free plan (Basic plan) | Free |
| 2. | Zoom Pro | $149.90 per year. |
| 3 | Zoom Business (for small and medium-sized businesses) | $199.90 per year |
| 4 | Zoom Enterprise (for large enterprises) | $199.90 per year. |

There are a few other add-ons (optional) plans you can choose to get, depending on your needs. These plans serve as compliments to any of the paid plans, as they give you access to more features, thereby making the Zoom experience better for your organization. They include;

| 1 | Audio plans | Starting at $1200 per year. |
| --- | --- | --- |
| 2 | Cloud storage | Starting at $480 per year. |
| 3 | Large meetings | Starting at $600 per year. |

# Chapter 3

## Getting Started With Zoom

## Downloading and Installing Zoom App

### Using Your Computer

1. Go to the official Zoom website at https://Zoom.us/download.

2. This will open up the download center to you. From the download center, click on the "download" button. You will find this option under "Zoom client for meetings."

3. Clicking on the button in the step above will cause the application to be automatically downloaded to your computer.

4. When the download is complete, click on "install" to launch the downloaded application and install it on your device. Installing the application gets it set up for use, so you can go ahead to follow the steps in the section below to sign up for the service.

## Using Your Mobile Device

1. Launch the Play Store app and wait for the home screen to load.

2. In the home screen, tap the search icon at the top of the screen and type in "Zoom." This will produce search results that will show on your screen. From the list of options that have come up on your screen, click on Zoom cloud meetings to open up its dedicated page.

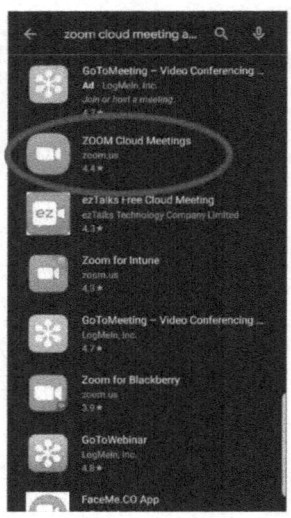

3. When the dedicated page has opened, you will see a big green button that has "install" written on it. Tap on this button to start the process of downloading the app to your android phone. This action will open up a

screen that shows you the list of applications and phone features that the Zoom app needs access to for it to function properly. Click on the "accept" button at the bottom of this page, and allow the application to install.

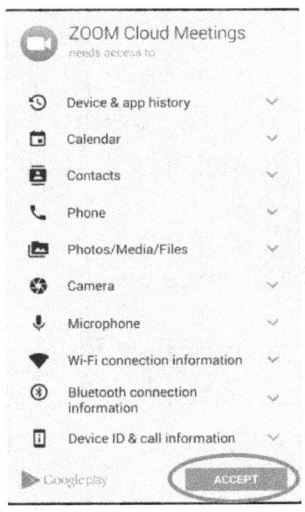

4. When the application has installed, launch it and follow the on-screen instructions to sign in to an account you own or follow the instructions in the section below to sign up for the Zoom service.

5. The processes for installing the application on your apple devices are pretty much the same. All you need to do to get started is to open up the Apple app store in your device and search for the Zoom application and

follow through with the onscreen prompts displayed on your device's screen.

## Signing Up For a Zoom Account

Signing up for a Zoom account is an easy task. You can get this done using any device that supports the Zoom app, including a mobile device, Windows or Mac device. To get this done, just follow the steps outlined below.

## Using Your Mobile Device

If you are using a mobile device (iPhone, iPad or Android device), the steps are pretty much the same across these three platforms.

1. Download the Zoom app from your device's application store (Google Play Store for Android devices and the Apple app store for Apple devices). Finding and installing this application from the store has been described earlier. All you need to do is open up the store app and search for Zoom. Click on install/download and wait for the process to be completed.

2. After downloading the app from the store, launch it. On the welcome page, you will see options to either "sign up" or "sign in." Tap on sign in. This will open you up to another screen where you can continue the sign-up process.

3. On the new page, enter the email address that you would like to use for the Zoom account. Make sure the email address you enter is a valid one and that you have access to it. This is because the platform will send a confirmation mail to this email, and you will need to access the mail to completely set up your account. Also, periodic updates and other notifications will be sent to your email address on a regular basis. You want to be sure that you have access to the email address in question, so you do not miss out on these vital notifications.

4. After entering your valid details, check the tiny box under that says, "I agree to the terms of service." To be at par with the terms of service in question, you can click on the "terms of service" characters on that page. This will open you to a page where the terms of service are detailed.

5. When you are done, click on "sign up" on the page. A pop-up notification will appear on the screen when you

have done this, telling you that a confirmation mail has been sent to the email address you entered. Tap "ok" to close this notification.

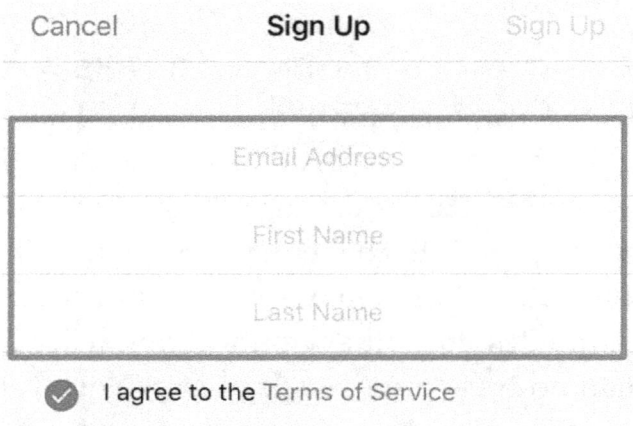

6. Open up the email you received from "Zoom" and click on the "activate account" button in the mail. When you have done this, the link will be opened in your phone's browser.

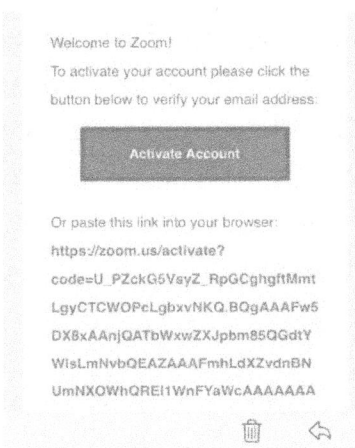

Welcome to Zoom!

To activate your account please click the
button below to verify your email address:

**Activate Account**

Or paste this link into your browser:

https://zoom.us/activate?

code=U_PZckG5VsyZ_RpGCghgftMmt

LgyCTCWOPcLgbxvNKQ.BQgAAAFw5

DX8xAAnjQATbWxwZXJpbm85QGdtY

WIsLmNvbQEAZAAAFmhLdXZvdnnBN

UmNXQWhQREl1WnFYaWcAAAAAAA

7. When this has opened up in your browser tab, begin the confirmation stage by making sure that all the details filled in are correct and valid. After doing this, choose a password. In the process of choosing a password, make sure that the password you choose is secure enough to protect your account, and one that you can remember because without your password, you will not have access to your Zoom account. The classic prerequisite for a password to be approved for use by Zoom is that it must contain 8 characters, and out of these 8 characters, at least one must be a letter or number. If you select a not acceptable password, instructions will appear on-screen to help you tweak your choice and select a password that is in sync with what the app accepts as passwords.

8. After you have selected a password, you can go ahead to invite colleagues or students to a meeting immediately. The next step after your password has been set successfully encourages you to start a meeting immediately. However, if you are not looking to do this, you can skip this step.

9. The step after this one encourages you to start a meeting immediately. If that is still not what you want to do, evade it by clicking on "go to my account." This will take you to a page where you can fill in the details you used to sign up for the service; your email address

and password. When you are done with this, click on "sign in." This gives you access to the account you just created and from there, you can get started with using the Zoom app.

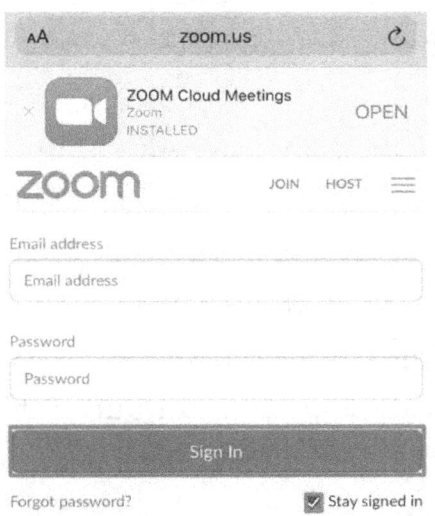

## Using Your Windows Browser

Follow the steps outlined below to sign up for Zoom using your Windows browser.

1. Type in the URL http://Zoom.us on your browser and click on the Sign-Up button

**Note:** Register with your email address, Facebook account or your Google account. Open your account with your business email address if it is meant to be for business purpose and if you have one. Also, ensure the email address provided is active because you will be sent an activation email.

2. After clicking the sign-up button, you will be requested to type in your date of birth.

3. When you are done with this, an activation email will be sent to the email address you provided, requesting you to activate your account. Click on the activate your account button to activate your account. If you don't receive the email in your inbox, check your spam folder and verify that the email address provided during registration is valid.

4. On the other hand, if the activate account button does not respond, there is a link typically sent alongside it; copy and paste the link into your web browser– this will also activate your account.

5. When the webpage opens, you will be asked if you are signing up for a school. Click yes, if you are and no, if you are not.
6. Afterward, you will be taken to another page to complete the registration process, such as your name and password.

   **Note:** Your password must contain the following:
   - Should be 8 characters
   - At least one number
   - At least one letter
   - Both upper and lower cases
7. Upon completing the previous step, you will be asked to invite people to create their Zoom accounts. Skip this step because it is not necessary for now.
8. Upon skipping, you will be provided with a URL that will navigate you to where you can start your meeting. Click on the "start meeting now" button to receive a prompt requesting you to download the Zoom desktop application. Follow the on-screen instructions to download and install the app.

9. When the installation is completed, two options would be displayed, "Join a Meeting" and "Sign In." Click on "Sign In" given you want to start a Zoom meeting.

10. Once you are signed in, go to "Home," and search for the "New Meeting" button, then click on it. With this, you are ready to start your first Zoom meeting.

## Configuring Secure Video and Audio Settings

Earlier versions of the Zoom app had one major flaw; spammers could join in for a meeting unannounced and uninvited. While the owners and developers of the platform did all they could to work a way around this, it did not change the fact that it was a very unwelcome demerit of making use of the application because once these people got access to the meeting rooms, they could go on to create mayhem in the meeting rooms.

Further updates to the application helped to curb these excesses. For one, a lot of methods were included in the application, just to make sure that the privacy of those that attend meetings was secure and that people with bad intent stopped getting uninvited access to meeting rooms. These updates to the Zoom app come with features to protect both the video and audio components of meetings.

Some of these updates include the following;

1. The meeting host can choose to change the audio settings. Just like how the meeting host controls what happens in a physical meeting, this feature was geared towards replicating that same scenario in an online meeting.

As the meeting host, you can choose to mute everyone or a particular person in a meeting. Muting ensures that no sound is captured from the mics of those that you have muted. Suppose you have integrated to your Zoom app a third-party website/tool. In that case, you can choose to require that participants of the meeting go through an orientation material that teaches them what is acceptable and the rules that guide communication in the meeting.

Conversely, if you feel it is not entirely necessary, you can choose to mute all the participants of the meeting from your end. If you do this as the meeting host, you have to be the one to unmute them for them to make use of their microphone. If you aren't going to do this, you can also ask them to use the chat section in the meeting room. These features reduce the chances of your meetings getting interrupted by spammers and uninvited guests.

2. As the meeting host, you can tweak the video settings before the meeting, i.e., to turn off the video component of those attending the meeting. Not many meeting participants love to show themselves during meetings with Zoom, and doing this is one sure way to make the participants of your meeting feel more comfortable. This also implies that if the participants of your meeting are comfortable with their videos being blocked, there is less likelihood for your meeting to be interrupted unnecessarily.

## Scheduling a Zoom Class or Meeting

One of Zoom's amazing features is that it allows you the luxury of scheduling meetings ahead of time and inviting those that must be a part of the meeting you have scheduled.

Scheduling meetings in Zoom is not a difficult task. To get this done, all you need to do is follow the processes outlined below;

1. When you have signed into the application, select the "Schedule" button on the home screen.

2. Taking that action will open you up to a new page. On the new page, you will get to fill in the details of the meeting you want to schedule. Details like the time of the meeting, duration, name of the meeting, and whether or not it will be a recurring meeting are all set on this page. While setting up this page, you want to make sure that you do not skip the time zone settings at the bottom of the page. Skipping the time zone settings can cause you to create a meeting that many people will not be able to attend because of differences in time zones.

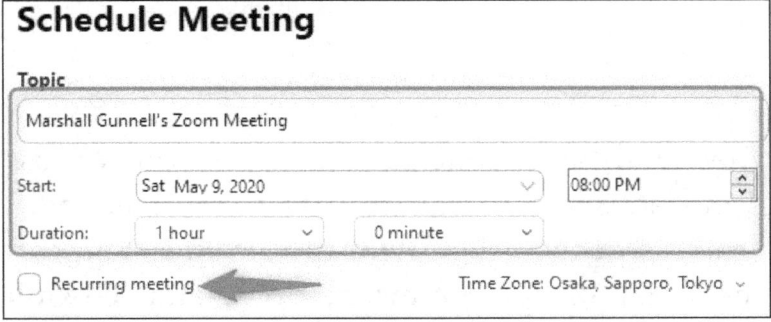

3. When you have finished with this step, you need to move on to set up the meeting ID. If you are having a personal meeting, you can go ahead to use a personal

meeting ID, but if you are looking to have a business meeting or a classroom lecture with many people being in attendance, you should select the option that allows you to "generate automatically."

4. Adjust the video and audio settings to what makes you feel better about the meeting, and what is required from the meeting. This option gives you some control over the meeting as the host, and you can toggle audio and video components on and off as you deem fit. In this step, you can also select how you want the participants to dial into the meeting.

**Video**
Host: ○ On ● Off          Participants: ○ On ● Off

**Audio**
○ Telephone          ○ Computer Audio          ● Telephone and Computer Audio

5. When you have done all these, you can choose to send out an invite to participants of the meeting, and you can also choose how you can do this. You can do this as a mail or link the meeting date and schedule to their Google calendars and Outlook tools. This way, you will not be worried about them remembering to attend the meeting or not, because the tools are automated to notify meeting participants and get them

to attend the meeting on the day it has been slated to hold. There are still a few customization settings that will come up after you have finished these stages. Follow them to add extra levels of security to the meeting that you have created.

**Starting a Zoom Class or Meeting**

1. Sign up for the service, and sign in to your account.

2. On the homepage, click on the "meetings" tab in the menu by the screen's left side. This will open up to you a list of meetings and classes that you have access to.

3. From the list that comes up, select the class you want to start/enter and click "start." Once you have done this, you will be taken to the meeting. Follow the prompts that come up on the screen to ensure you are not denied from accessing the class you want to access.

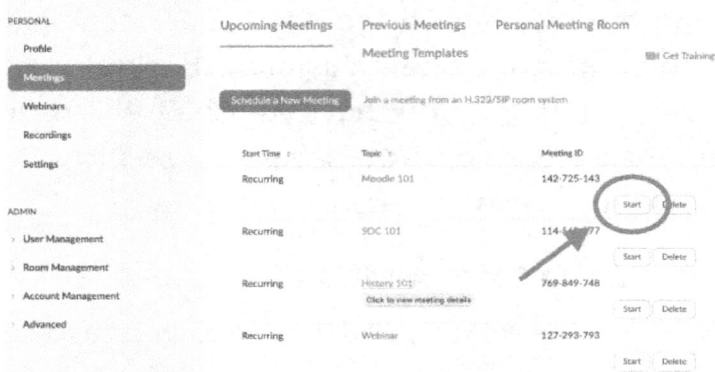

If, on the other hand, the meeting you want to access has not yet been created, you can do so by following the steps listed in the section above.

Start from signing up for the service, then follow the instructions that have been discussed earlier. The process is as easy as whatever you can think of.

### Inviting Students or Participants to Join a Meeting

Now that you have created a meeting, it is time for you to invite your students or other participants to be a part of the meeting.

### With Your Mobile Device

1. Sign in to your Zoom account on your mobile device. On the "meet and chat" page, tap on "schedule" a

meeting. This will take you through the processes that were explained in two sections above; how to schedule a Zoom meeting. Follow all the steps in that section, and be sure to keep up with the on-screen instructions.

2. When you are done with setting up the meeting, selecting the ID you would love to use for the meeting, and adding a password (if you so wish to), you will be opened up to another page where you can get started with adding invitees for the meeting. These are the people that will get a notification that the host (you) is inviting them to a meeting, so you have to make sure that the details you are adding here are correct, and that with them, you can be able to lay hold on the invitees.

3. At the top of this page, you will see an "add" button. Tap on the button to add invitees, and depending on the settings of your account/device, you may see a pop-up that requests you to give "Zoom" access to your contacts. Accept this pop-up, then tap "done."

4. This will completely create the event. Open up the meeting you just created and take a look at the top of the page. On that side of the page, you will see a button that asks you to 'send invitations" to people that should be in the meeting. This is where you get started with the

process of getting as many people as should be in the meeting to the venue.

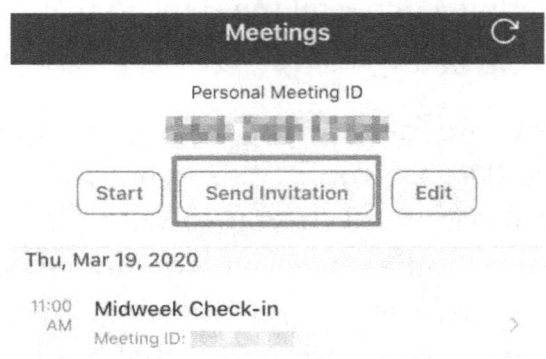

5. Tapping on the button in the number above will open you up to a screen where you can choose the people you want to send the invitation to, and how you want them to receive the invitation you have sent them. You can choose to send the invitation through email, or as a text message. It is best that you have the invitees in mind. Send the invitations using a means that they will be more likely to access immediately and join the meeting. Conversely, you can copy the link to the meeting and paste it anywhere that allows for copied content to be pasted on it.

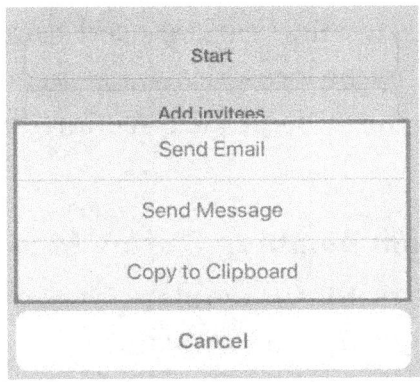

Start

Add invitees

Send Email

Send Message

Copy to Clipboard

Cancel

6. Click on the preferred option to open up a new tab where you will get to fill in the contact details of the person you want to reach with the invitation link. Whoever receives the invite from you will see that you are inviting him for a training session, and the notification will contain a link to have access to the class.

**With Your Computer**

1. Open up the desktop app and click on the "meetings" tab at the screen's top corner.

2. Follow the processes discussed in scheduling a meeting to fill up the required details and know what action to take. Once you have successfully scheduled the meeting, click the "copy the invitation" button at the

page's bottom. This link can be used to reach out to the people who must be present in the meeting, using any of the methods described in the section above.

## Preventing Zoom Bombing During Meetings

There are always those people who wouldn't want to behave as though they are uncultured people. They are everywhere, and you'd most likely find them online. Zoom bombing is the unwanted, disruptive intrusion, generally by internet trolls, into a video conference call. Remember that in earlier parts of this book, we talked about the fact that there could be those times when people wander into your Zoom meetings uninvited, and they are said to have started Zoom bombing if they intentionally put up an act that disrupts the quiet pace of the meeting.

As a meeting host or an instructor that hosts classes using the Zoom platform, it is imperative that you learn how you can be able to immune your meetings from these people who have no great intentions for it. Proofing your meeting from Zoom bombers is extremely important and considering the fact that you only need to carry out a few intentional and little acts to keep this trouble at bay, it is interesting that many meeting hosts begin meetings without finding these holes and plugging them.

That is the whole idea behind this section of the book; to help you with a few tricks that can help you prevent Zoom meeting bombing. Here are a few ideas to help you with this;

1. As much as it lies within you, refrain from making use of your personal meeting ID when creating new Zoom meeting rooms. You may want to think of your meeting ID as your phone number and a line of direct access to you. If you are to make use of this ID every time, and you are not careful about who has access to it, Zoom bombers are more likely to get hold of it, and with it, they are more likely to guess their way into your meetings and create troubles for you as the meetings unfold.

Alongside this point, never post your personal meeting ID at a place where it is public. Doing this is almost equivalent to handing olive branches over to Zoom meeting bombers, and inviting them over for a nice little chat during your next meeting (the only challenge being that they hardly ever have a nice little chat).

2. Always password your meetings, and implore every one that has access to the meetings and passwords not to hand over the passwords to other people. Your meeting password is the first line of security as without it, no one can have access to the meeting room. Creating a Zoom meeting and launching it without a password is almost the same as creating a physical room for a meeting with top executives of a firm, and not being careful enough to put a lock in the door so that only selected people will be able to attend the meeting. It just does not add up.

3. Maximize the use of waiting rooms. As the host of a meeting, you want to make sure that all the people who attend your meeting are approved as people who should be a part of the meeting. If you do not maximize this feature, you may not be able to achieve this aim. The waiting room is a feature that allows you to vet and approve all the people who want to join a meeting. When someone comes across the link to your meeting and clicks on it to join the meeting room, he is redirected to the waiting room first, and you will have to approve him from the backend for him to have access to the meeting room. This is another amazing feature that Zoom has added to their platform to make sure

they give you more control over what happens in your meeting rooms.

4. One of the major reasons people are able to Zoom bomb is that the meeting host has access to more privileges than they needed to have in the meeting. Those who are unable to manage the access they have will tend to abuse it. It is your duty as the meeting host to nip this tendency in the bud from the start. Here are a few ideas to put in place for your next Zoom meeting if you are going to prevent Zoom bombing;

- Disable the audio and video components of the meeting attendees from your end. If attendees discover that you have left their mics on, and that their live video feeds are on, there is every tendency that they will begin to stream without taking permission from you.
- Turn off screen sharing for everyone, except the hosts of the meeting. The fact that people have the ability to share their screens using Zoom does not mean that everyone in a meeting room should do the same. This feature should only be available to the anchors of the meeting. If people are able to

share their screens unrestricted, that will be a lot of interruptions to the flow of the meeting.

5. You may also want to lock a meeting as soon as it starts. Although this will most likely keep away the latecomers (at least for a while), the hassles involved with having to deal with Zoom bombers and trying to eject them during the middle of a meeting is worse than putting this little measure in place from the start.

6. Consider using an invite-only meeting. This option means that only those that received an invitation from you with a link can have access to the meeting/class. This way, you get to reduce the number of people who stumble into the meeting. While at it, encourage those you send the invitation to avoid sending it off to other people without gaining permission from the meeting host first.

## Recording a Meeting

This is one of the most innovative features of the Zoom app, as it allows moments from the meeting to be captured and preserved in a retrieval/storage system. This feature is great for making sure that people can relate to previous meetings and get the most they can get out of meetings.

In the introductory part of this book, different types of plans available for purchase on the Zoom app were mentioned, and these plans were explained with respect to their unique features. How much you can record a meeting depends on the chosen plan, and to be more clarified, please refer to the section of this book where Zoom plans were discussed in more detail.

Recording a meeting in Zoom, just like every other thing, is not as difficult as it may seem. Here are the steps you need to follow to get this done;

1. Set up the meeting and have the participants join the meeting. They need to be present so that you can officially start the meeting and begin to record as the meeting progresses.

2. Once the meeting is in full swing, start recording by clicking the "record" button on your device's screen. Conversely, you can start a recording by clicking Alt+R simultaneously.

3. Selecting the record button will automatically begin the process.

However, to pause/stop the recording, make use of the "pause/end recording" feature on your screen. Note that these icons do not appear until you have started recording. Once you have started recording, you will be able to see them on your screen. When your meeting is over, or when you are done with recording, you can click the stop recording button, and then end the meeting.

4. Ending the recording will prompt you to save what you have recorded to a storage device. If you are using the paid version of Zoom, you will have access to some storage space over the cloud but if you are using the

free version, you have to sort these details out by yourself.

## Grant Participants the Ability to Record

Inasmuch as Zoom allows for the participants of a meeting to record the meeting, this feature is only available to the meeting's host. This is another feature put in place by the Zoom team to ensure that the host has the most control over the meeting room.

However, even as the host, you are allowed to grant attendees the power to record a meeting/class. This may be because of a lot of reasons, including your desire to focus on delivering your lecture/presentation and not have to worry about technical issues with the meeting room. Whatever the case, here's how you can allow an attendee/student to record a meeting while it is ongoing.

1. When the video conference has started, select the "manage participants" tab at the bottom of the screen. This will open up to you the list of all the people present in the meeting at the time.

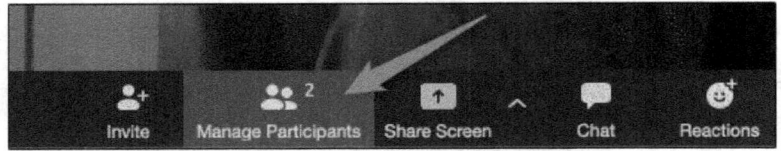

2. By the name of the participant(s) you want to give access to record the class, you will find a "more" button. Click on this button, a drop-down menu will appear afterward.

3. In the menu, select the option that allows the participant to record. This is the "allow record" button. Clicking this will give the participant(s) the ability to record the meeting/class as it goes on.

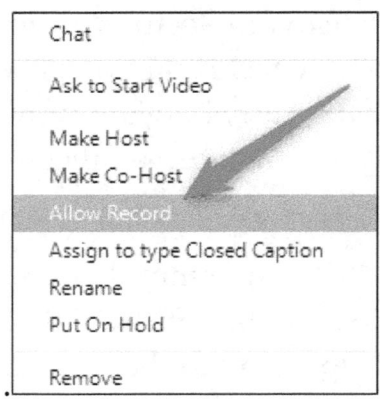

**Sharing Your Screen**

This is another feature of the Zoom platform that fosters professionalism and spot-on presentations. With the host's ability to share his screen with the attendees or students of a meeting/class, he would be able to deliver his presentation as though they were congregated in a place. If you are going to make highly impactful presentations, you must share your screen as a teacher.

To achieve this, just follow the steps outlined below;

1. Log in to the meeting room and start the meeting with the attendees. In the meeting room's home screen, click on the "Share screen" icon to begin sharing your screen at any time of the call.

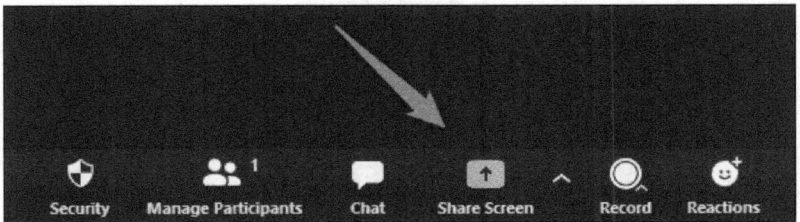

2. Conversely, you can make use of the shortcut keys; Alt+S on Windows 10, and Command + Shift + S if you are making use of a Mac device.

3. Going on with steps 1 and 2 will open up to you another screen. This is the "basic" tab of the share screen options. From this screen, you can choose which screen you want to share (this is only applicable if you are connected to many monitors/tabs, if there are other supported apps that are running in the background, or if a whiteboard is open somewhere. In this screen, you have to choose which screen you want to share with the meeting attendees, and when you are done, click "share" at the bottom of the screen.

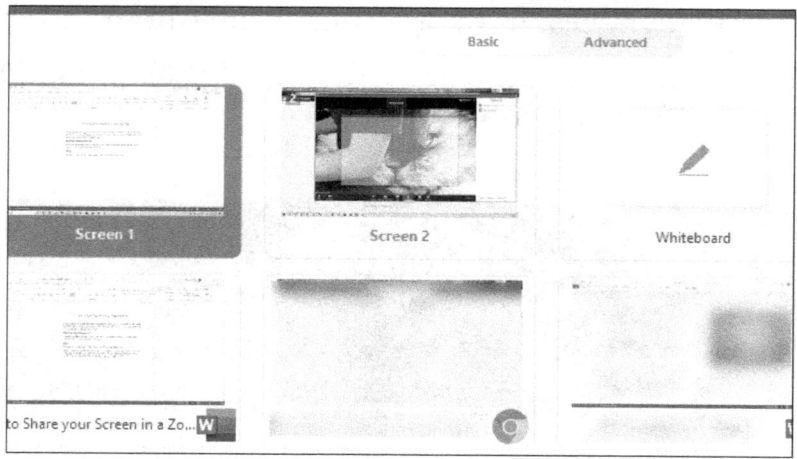

4. When you are done with screen sharing and would love to toggle the feature off, click on the red "stop share" button that is visible on the screen you are

sharing. This will immediately end the screen-sharing and take you back to whatever you were showing the students/meeting attendees before you began sharing your screen.

## Allowing Students/Attendees Share their Screen

Inasmuch as you are the meeting host, if there is need for you to allow other people in the meeting room to share their screen, there is a feature ingrained in the Zoom app that allows you to achieve just that. However, it is vital that in a bid to maintain the sanity of the meeting room, you should make it a point of duty to use his feature sparingly, and only when it is entirely necessary.

Follow these steps to achieve this;

1. The meeting needs to be ongoing, and the participant you want to share his screen has to be present in the meeting room and ready to share as you give him permission to do so.

2. During the meeting, you will find the screen share option on the screen. Next to this button is a small arrow. Select the arrow from the menu that will appear afterward and select the "advanced sharing options."

3. When the advanced sharing window has opened up, you can go on to select who can share their screens, when they can share their screens, and how many people are permitted to share their screens per time.

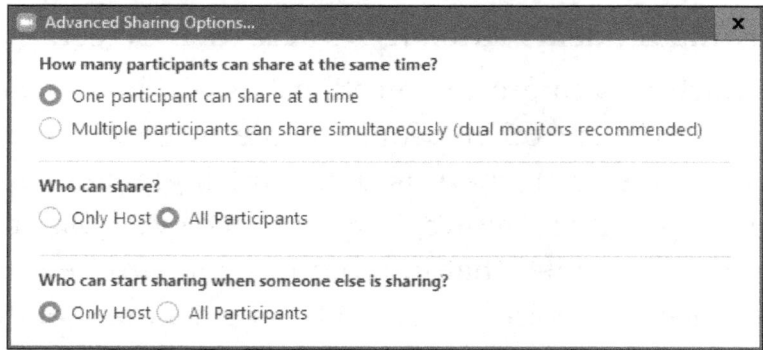

4. When you are done with selecting these options, click on "done" to activate the settings you have enabled. Notify the person you have granted permission and get him to set up from his end so that he can start sharing his screen from his end.

Although you have given someone the ability to share his screen, you are still the meeting host, which also gives you the ability to stop his screen sharing. These features have been included in the platform to enable and foster collaboration while still maintaining an atmosphere of decorum among the meeting/class participants.

**Setting Up Zoom Whiteboards**

In the beginning part of this book, we established the fact that the Zoom app was built with collaborations and a seamless virtual meeting experience in mind. The whiteboard feature is one of those tools that fosters collaborative efforts among team members and ensures that they work together as though they were physically congregated, even though they may be worlds apart.

A whiteboard is a built-in feature of the updated versions of Zoom that allows participants in a meeting room to collaborate on projects, find solutions together, and brainstorm ideas using visual aids like whiteboards. This tool allows you to share whiteboards and get to annotate them with other people in the meeting room at the same time. The best part of this feature is that it is available to all Zoom users - both paid and free users.

This is how to create and make use of whiteboards in meeting rooms.

1. Log in to your Zoom account and start up a meeting.

2. In the meeting's home screen, you will find the "share screen" at the bottom of the page. This will open up another menu for you. The screen that will pop up will contain options related to screen sharing; advanced screen sharing options.

3. From the options displayed, select "whiteboard." Once you have selected the whiteboard option, another screen will open up, asking you to customize the whiteboard user experience. This is where you select how you want the members of the meeting room to interact with the whiteboard you have created, including whether they can use text, draw, use preloaded shapes on the board, or create theirs on the go. After ensuring that you have effected all the settings, click on the button that allows you to save.

This publishes the button that allows everyone in the meeting room have access to the whiteboard. Before publishing your whiteboard, you can choose to save the whiteboard to use the same in the future as a template.

4. Publishing the whiteboard in the meeting room does not mean that the meeting participants automatically get to contribute to the board. If you want to make sure that the people in the room get to contribute to the board, you have to allow them to do so. To achieve this,

click on the "more" button (the three dots at the top end of the whiteboard screen). This will open up a menu from where you can select that you are granting everyone in the room the permission to make changes to the board as the meeting unfolds. From this menu, you can also choose to enable/disable annotations from the meeting participants.

5. When you are done with using the whiteboard, click on "stop share" button in the meeting's control panel. This will take all the participants' screens back to what it was before you shared the whiteboard.

## Setting Up Zoom Annotations

Annotations are features that are used together with the whiteboard feature, and they make the collaboration game easier and more interactive. You can decide to turn on this feature to allow every person in the meeting room to contribute to the whiteboard at the same time or choose to disable annotations. This way, the meeting attendees can only get to see what you are doing and listen to you while you work, but they do not have the permission to make direct changes to the whiteboard in view.

During a Zoom meeting (or in preparation for the meeting), annotations can be enabled for making direct changes on the whiteboard in the meeting room. For annotations to be used in the meeting room, you must (as the meeting host), enable them from your end. This is how you can get this done;

## Using Annotations As a Free User

1. Sign into your Zoom account and go to settings. From the settings tab, find and click on meetings."

2. This will open up the basic meetings settings tab on your device. From the list of options that appear on your screen, find and toggle on "annotations." if annotations were disabled before, you will get a drop-down verification dialog, asking you to confirm whether or not you want to save the change that you have effected. Confirm that this is what you want to do to save the change.

**Annotation**
Allow participants to use annotation tools to add information to shared screens ⓥ

**Whiteboard**
Allow participants to share whiteboard during a meeting ⓥ

○ Auto save whiteboard content when sharing is stopped

## Using Annotations As a Paid User

1. Sign in to your paid Zoom account, and find the "account management" setting.

2. Click on "account settings," and from there, click on the "meeting" tab. This will open up to you the basic meetings tab.

3. From the tab that has been opened, scroll down and toggle on annotations. You can decide to toggle the button on or off, depending on your needs.

## Accessing Annotation Tools

Now that you have toggled on annotations, the next thing you need to do is to be able to access the tools. There are quite a number of them. To access these tools, do the following.

## Using Your PC

1. Pull up the whiteboard that you want to work with.

2. You will find the annotation tools carefully laid out at the bottom of your screen. Click on any of these icons to make use of any of the tools displayed.

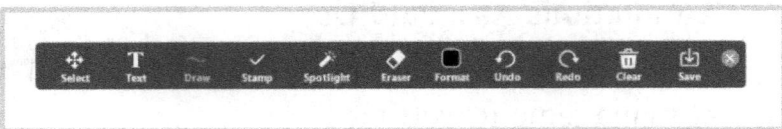

3. You can also annotate on any of the screens you are sharing. The people in the meeting can also see what you are doing as you share your screen and annotate on them. To get this done, simply click on the screen you want to share, follow the steps described in the section above (on how to share your screen), and click the "annotate" button from the toolbar floating on that screen.

## Using Your Mobile Device

The options available on your mobile device are not as expansive as what you get if you were to use the desktop version. To annotate on a shared screen using your mobile device,

1. Log in to your Zoom account and start a meeting.

2. From the meeting home screen, tap on the "share" button at the screen's bottom.

3. Select "screen" from the list of options that will be shown to you. You will need to click the box that says "you are granting Zoom the permission to record potentially sensitive content."

4. Once these are in place, go back to the screen you are sharing (the whiteboard in this context), and tap on the little "annotate" button at the bottom of the screen. This will give you access to all the annotation tools available in the mobile version of the platform.

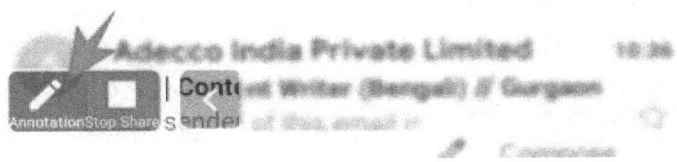

## Setting Up Breakout Rooms

Breakout rooms give meeting participants more room to experience collaboration on a new level. They are smaller rooms where all the people who gain access have the task of interacting closely on issues of concern, sharing ideas, brainstorming, or even getting to learn on a more personal level.

The breakout rooms feature allows you (as the meeting host) to split your general meeting room into 50 different rooms. To make use of breakout rooms in the meeting, you need to first activate breakout rooms in the meeting set up, and you can also pre-assign meeting participants to breakout rooms. This way, they can easily toggle between their breakout room and the main room.

### Creating Breakout Rooms

1. Sign in to the platform and start a meeting.
2. Click on "breakout rooms."

3. Select the number of breakout rooms you want to create and indicate how you want the meeting members to be assigned to the breakout rooms. You have the option of assigning participants automatically (this way, Zoom splits the participants of the meeting evenly across the number of breakout rooms you have created), manually (you select who goes where), or you can allow the participants of the meeting to make their choice.

4. Click the "create breakout rooms" button. This will create the rooms based on the options you have selected. Creating breakout rooms does not start them automatically.

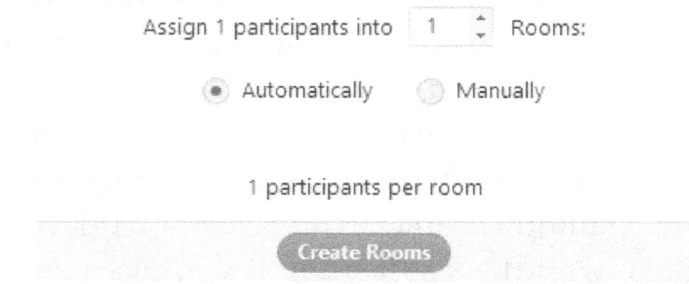

5. When you have created your breakout room, click on "options" to view the extra options available to your created rooms.

6. From the list of options, select anyone that applies to the room you created. The options available are shown below;

- Allow participants to choose their rooms themselves.
- Allow the participants to toggle between breakout rooms and the main meeting room as they wish. This is handy if they will have a need to move to and fro.
- Automatically move assigned participants into breakout rooms. This option will activate at the right time, and everyone that you have assigned to rooms will be moved into the room so long they are still in the meeting room.

- Auto close breakout rooms after () minutes. This is where you get to choose how long you want people to remain in breakout rooms. When the time elapses, breakout rooms are closed and people are returned to the main meeting room.
- Set countdown timer. This usually work hand in glove with the last feature. If you place a peg on how long you are allowing people to remain in breakout rooms and click this button, participants of the meeting will see a countdown timer once they get into a meeting room. This is necessary to make sure that they stick to time and not get sidetracked in breakout rooms.

**Assigning Participants to Breakout Rooms**

As stated earlier, you can create breakout rooms and assign participants or students in there. This way, they do not get to choose which rooms they want to be in, but at the right time, they are all moved into the room you have assigned to them.

To assign participants to rooms

1. Mouse over to the breakout room you want to assign participants to and click on the "assign" button on the breakout room card.

2. Select the participants you wish to add to the room and go through this process for all the created rooms. When you assign a participant to a room, the available number of participants will automatically show next to the assign button.

**Managing Breakout Rooms**

As the meeting host, you can make changes to breakout rooms, even as breakout sessions are in progress. Although it is best to effect these settings before breakout sessions begin, that does not take away the fact that you can do this at any time.

Here's how to prepare and manage the breakout rooms that you have created

1. Mouse over to the breakout room you want to effect changes in and click on the name of one of the people in the room. This will open up a list of actions that you can take with that person in the room. You can choose to;

- Move the participant to another room.
- Exchange the selected participant in one room with another participant in another room. This will swap their positions in rooms.

2. In the breakout room, you can choose to delete it. You can do this to bring the session going on in the breakout room to a stop, and taking this action will move all the people in the room to the main meeting room.

3. Another thing you can do as the host of a meeting is to add a new room. This option creates a new room that you can customize immediately.

4. As the host, you can also choose to

- Join a breakout room in session. This way, you can swap between rooms and keep tabs with what is going on in them.
- Leave a breakout room.
- Close all rooms. When you close all rooms, all the rooms' participants get a notification that they have a limited amount of time to pull themselves together because they are about to be taken back to the main room and have their breakout sessions stopped from your end.

**Sending Broadcast Message to Breakout Rooms**

1. Go over to the meeting controls panel, and click on the "breakout rooms."

2. Click on "broadcast a message to all." this action will open up a page for you.

3. In the page that has opened up, type the message you want to send to the breakout rooms. Note that whatever you type here can be seen and read by everyone in the different breakout rooms, so you may want to be careful with the information you pass across.

4. When you have typed in the message you want to send to the rooms' participants, click on "broadcast" and the message will be displayed for all the people in the different rooms to see.

**Setting Up Polling For Students/Participants**

Polls are a great way to gather participants' feedback in a heartbeat. The great thing is that the Zoom platform allows you to create, send and even gather results from polls almost immediately. This way, you can get to see what the participants of your meetings think about salient topics, and if done well, you can use the results from your polls to improve your meetings going forward.

Here are a few things to know about polling;

1. You must be a licensed user of the platform to create a poll.

2. Polls can only be created for scheduled meetings. If you must create polls for non-scheduled meetings, then you need to use your Personal Meeting ID as you get ready for the meeting. When using your personal ID, remember the best practices discussed earlier. If not, please make use of scheduled meetings for your polls.

This is how to enable polling for your Zoom meetings

1. Sign in to your account and select "account settings." You will find this under the "account management" section of the administrator control panel.

2. In the meeting tab, scroll down and seek out the "polling" option towards the lower half of the screen. Toggle the activation button to enable polling in the meeting you are about to create. Now that you have enabled polling, it is time to create the poll.

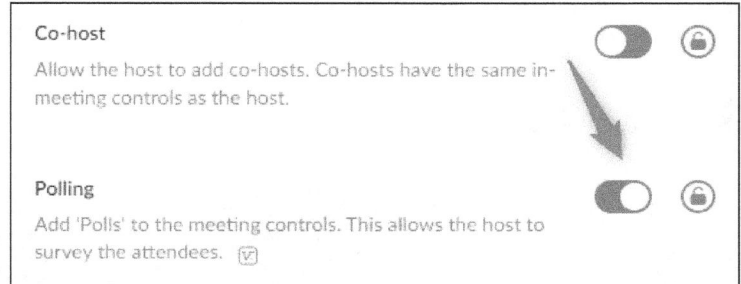

**Co-host**

Allow the host to add co-hosts. Co-hosts have the same in-meeting controls as the host.

**Polling**

Add 'Polls' to the meeting controls. This allows the host to survey the attendees.

3. Go back to the home screen and select the "meetings" tab from the left-hand control panel.

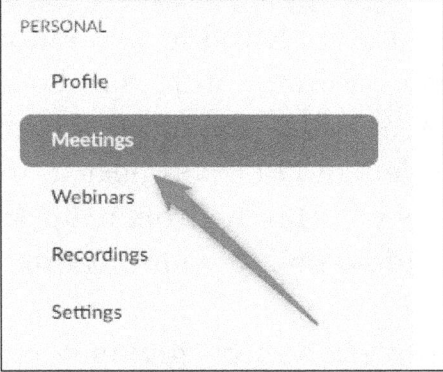

PERSONAL

Profile

**Meetings**

Webinars

Recordings

Settings

4. Go through the motions of setting up and scheduling a new meeting described in earlier sections of this book. You can also select an already scheduled meeting from your list of scheduled meetings if you are not looking to create an entirely new meeting.

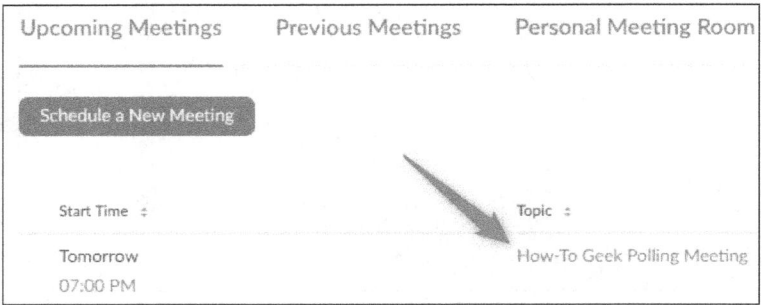

5. When you have opened up the scheduled meeting, you want to add a poll (or create a new scheduled meeting to add a poll). Scroll to the bottom of the page and you will find a little notification box that tells you that you are yet to add a poll to the scheduled meeting. Click on the little "add" button next to this note at the bottom of the page. This will open up to the window where you can add a poll to your meeting.

6. Start the setup process by giving your poll a name and deciding whether you want answers to be collected anonymously. If you accept to collect answers anonymously, you will only see the answers given back to you, with a "guest" beside it as the name of the person who gave the answer. If you want to know who gave what answers, you shouldn't activate this button.

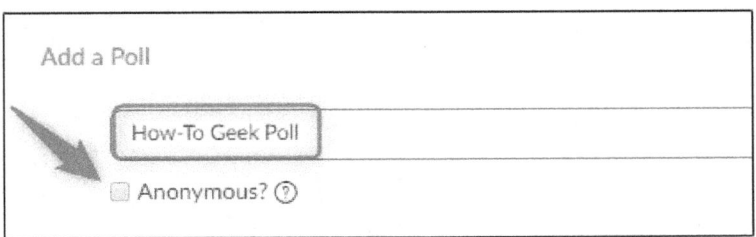

Add a Poll

How-To Geek Poll

☐ Anonymous? ⑦

7. After doing this, type in the poll question. You have only 255 characters, so you must make them count. When you have done these, select whether or not you want the poll question to be a multiple-choice question or the question that the students will give a single answer to. After this, type in the available answers you want to provide them with as options. Note that you can get to build up to 10 answers for the students to select from, but it is best you stick within a reasonable number so that the polls do not feel overwhelming to those who have to answer them.

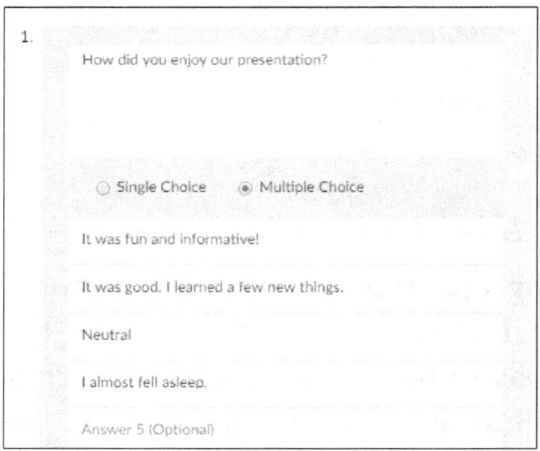

8. You can always add new questions to the poll by selecting the "add a question" button at the window's bottom. This opens you up to a place where you can go through the motions of filling in a new polling question and customizing the answer experience for the students that will be in the meeting room when the time is right.

9. Click "save" once you are done with the creation of the polls.

## Delivering Your Polls During the Meeting

Now that you have created your poll and it is time for the meeting, you can deliver the polls to the people present in the meeting.

1. When you are ready to launch the poll (during the Zoom meeting), select the "polls button at the bottom of the meeting window.

2. This will open up to you the window containing the polls you created. The essence of this is to get to review the polls you have created before letting it go live. When you have gone through the poll questions and are satisfied that they are just the way you wanted, click on the "launch poll" button at the bottom of the screen.

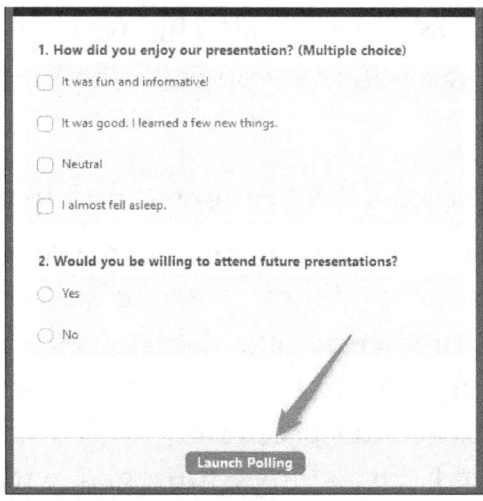

3. As people begin to give answers and the results start coming in, you will start seeing them from your backend. To speed up the process, send out a notification to those in the meeting room and request

that they attend to the polls as quickly as they can. Once everyone is done with the voting, click on "end polling" at the screen's bottom.

4. This action will show you the polling results and you can choose to share the results with the students in the meeting room or relaunch the polling for them as the case may be.

5. If you want to return and view the polling results again, you can find them in the "poll report" option of the "previous meetings" tab. This will open up to you the results of the poll that you collected in the meeting.

## Setting Up Quizzes For Students/Participants

Another interesting feature of the Zoom app is that it has a feature that allows users to set up their own quizzes and interact among themselves with the easy-to-use and very handy tool. With this feature, you can set up a quiz, get the people in a meeting room ready and interested in going through with the quiz immediately, and compute results simultaneously. While setting up the quiz, you must have a clear understanding of the quiz you want to set up and the number of rounds available during the quiz. Each round of a quiz can take up to 10 questions, and depending on the nature of the quiz, you can make as many rounds as you think is necessary for your needs.

## How to Set Up a Quiz in Zoom

1. Sign in to your Zoom account and set up a meeting.

2. Add the people who need to be in the meeting, or invite them to join using the methods discussed in earlier sections of this book.

3. By this time, you must have planned out your quiz, such as what questions you will ask those in the rooms to answer. It would help if you got all of your questions ready so that you can fire them away once the time is right.

4. Once you are done with the questions you want to ask, present them in a visual presentation. You can leverage the power of any of the visual presentation tools out there to do this; Powerpoint, Slides, Keynote, or any other application of this nature. The aim of presenting them this way is that they will be more likely to grab and hold the attention of all that are in the room. Also, the good part of making use of presentations is that you can get anything to be a part of the presentation; images, videos, tables, and infographics

can all be added to the presentation slides of quiz questions.

5. Prepare the answers to these questions while you are at it. You do not want to be overwhelmed when all the people's responses begin to get thrown at you.

6. Now that you have set up your quiz, send it off, alongside the meeting room's link, to all the meeting participants. They need to access it from the start to know what they are up to as far as the meeting is concerned. You can choose to do this by adding a reminder in the mail that they get to take an assessment within the day's class.

7. Once the meeting has started and the people are all settled in, start sharing the presentation you created. Make use of the screen sharing procedures (as described in earlier sections) to make this happen.

8. Turn off annotations and put every other person on mute. They should be able to feel the power of the quiz, just as they would have if they were to be in a physical classroom.

9. When the quiz is over, one of the easiest ways to mark the quiz and figure out the winners immediately is to have the meeting participants mark their own

quizzes. This is why you had to create the answers too from the start and compile them as a presentation. Just the way you displayed the questions, display the answers for them to assess their work and grade themselves. One thing to note is that this may be a good time to turn on the video and audio components of the meeting for all of the participants. This is necessary for you to keep your eyes on them and give them those feelings they are being watched. Allow them to have fun with the grading process and once they are done, you can get them to wave their scripts before the camera so you can see just how much they were able to get.

Alternatively, you can setup quizzes using the poll feature. Simply refer to the section on setting up and using polling above.

## Using Zoom Chat Tool

The chat tool is available for everyone in a Zoom class or meeting. This is one of the easiest and least distracting ways to get your thoughts and ideas out in public. With this feature, people can send direct messages to the meeting room, adjust the chat messages' settings to allow everyone in the meeting room to see the messages, or just allow the host of the meeting to see the messages.

**With Your PC**

1. Log in to your Zoom account and make sure that you are in a meeting. You cannot make use of the chat tool if you are not in an ongoing meeting.

2. At the meeting controls bar at the bottom of the meeting window, click on the "chat" button. This will open up the chat window at the right-hand side of the screen.

3. Type a message in the message box and click "send." This will send the message to all the people in the meeting room.

4. If you want to send the message to any particular person or group of persons in the room, click on the drop-down arrow next to the message box. This will open up to you the people in the room from where you can select who to send the message to.

## With Your Mobile Device

1. Make sure you have the Zoom app installed for your device. You can find them on the application stores for your respective devices.

2. Launch the app and enter a meeting.

3. Tap on "participants" on the control bar of the screen. This will open up a list of activities that you can carry out.

4. Tap on "chat" from the list that just dropped down. From here, you can craft a message to be sent to the group of people in the meeting, or follow the same procedure already discussed above to send the message to a selected few.

## Using the Chat Feature As the Meeting Host

Like every other feature of the Zoom app, the host has control over the app's chat feature, to a very large extent. You can choose to disable the feature so that no one else in the room can use it or enable it as the case may be. You can also restrict how participants can send messages to themselves by tweaking the settings to ensure that participants only get to send messages to you or a group of panelists in the meeting venue.

This is how you can go about it.

1. Log in to your account and start up a meeting. Have the invitees in the room and make sure that the meeting is underway.

2. Click on the "Chat" button in the meeting window (if you are making use of the desktop application), or "participants" then "chat" (if you are making use of the web version).

3. In the message window, click on "more." This will open up to you a list of options that you can get to choose from. The options vary from allowing the participants to chat with themselves, with yourself, or with no one at all.

4. If you are hosting a webinar, you can allow the participants to chat with themselves, with you, with no one, or with a set of panelists that includes yourself.

## A Short message from the Author:

Hey, I hope you are enjoying the book? I would love to hear your thoughts!

Many readers do not know how hard reviews are to come by and how much they help an author.

I would be incredibly grateful if you could take just 60 seconds to write a short review on Amazon, even if it is a few sentences!

>> Click here to leave a quick review

Thanks for the time taken to share your thoughts!

## Activating Non-verbal and Verbal Feedback

### How to Enable Non-verbal Feedback

As a meeting host, you can activate this feature for all the participants of your meeting. To do this,

1. Sign in to your Zoom account and click on "account settings.

2. From the tab that opens up, click on "meetings." This should open you up to the basic meetings tab.

3. From the drop-down menu that pops up, toggle the non-verbal feedback feature on or off as the case may be. Effecting a change to the status will bring up a confirmation box. Reinforce the fact that you are sure of what you want to do, and the setting will be saved for the meeting you are about to hold, and for other meetings you will hold moving forward. If you are looking to make this option mandatory for all the meeting participants, click on the lock (key) icon seen on the screen. Enabling this setting will ensure that participants of a meeting cannot boycott the setting you have put in place for the meeting, and it is quite necessary to make sure that decorum is upheld during the meeting.

**Sending Non-verbal Feedback**

1. Enter a meeting or class as a participant.

2. Click the "participant" button. This is usually found at the bottom of the meeting window. Clicking on this icon will open up to you a list of the non-verbal responses you can give the host of the meeting. It is vital that as a meeting guest or student, you try to engage the meeting this way because it reduces confusion and distractions in the meetings, while ensuring that the host can still interact with you in real-time. Note, however, that you can only activate one button at a time. So if one is active and would want to activate another, you have to start by deactivating the button that was active.

3. Here are some of the non-verbal feedback you can give the host during a meeting. You can choose to;
   - Raise your hand.
   - Indicate a "yes" or a "no."
   - Ask the meeting host to go slower or go faster.
   - Click the details button to get access to more non-verbal communication features available in the meeting window.

## Receiving Verbal Feedback

1. While setting up the meeting, you had the chance to mute or unmute the participants' mics.

2. To get verbal feedback, unmute the participants' microphones from your end, and encourage them to do the same from their ends as well.

3. It is best if anyone who wants to speak uses the "raise hand" feature first to permit him to, which will also help prevent rowdiness and unnecessary noise in the meeting room.

4. Once you have granted the person the opportunity, he can go ahead and speak into his microphone, audible enough for other participants in the room to hear.

## Transcribing Zoom Recordings

This feature allows you to have a transcript of your Zoom recording created for you and made accessible to you. Transcribing Zoom recordings is a great option if you are going to be making meetings available to people with special needs like those with hearing difficulty. Also, recording a meeting and having the transcript made available is quite beneficial because you can send the transcript over to the meeting participants much later on. With the knowledge that you will be sending them a transcript, they can focus on listening to you during the meeting rather than being torn between taking notes with the time and energy they should have spent giving you their attention.

Transcribing, however, is only available for users with the business, education, or enterprise license, and whose accounts have cloud recording feature enabled on them (in simple terms, those that are using Zoom on a subscription plan).

This is how you can get transcripts of your Zoom meetings;

1. Start a meeting and invite the participants.

2. Go through the motions of recording the meeting as discussed in earlier sections of this book (how to record your Zoom meeting).

3. To transcribe the recording of your meeting, you have three options available for you.

**A. Enabling transcription for your account**

I. Sign in to the Zoom portal and click on "account management" in the navigation menu. From there, seek out and click on "account settings."

II. From the options that open up to you, move over to the "recording" tab and click on the "cloud recording" button. Make sure that this feature is enabled on the account. If it is not enabled, turn it on manually.

III. Head over to the "advanced cloud recording settings" and click on the "audio transcript" checkbox to ensure that the feature is enabled. To confirm that this is the change you want to save to the account, click on "save." This automatically means that you have saved this change to the account. If you think you do not want it again any time in the future, you can come back here and toggle the settings off.

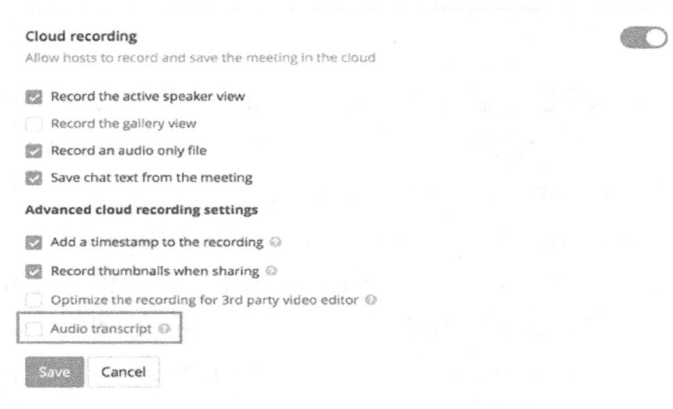

## B. Enabling transcription for your own use

I. Sign in to the Zoom platform.

II. In the meetings window, click on "settings."

III. This will open up a tab for you. In this tab, find the "recording" button and tap on it. From there, tap on "cloud recording" and ensure that the setting is enabled. If it is not enabled, toggle the button on and save the setting you have enabled. If you discover that the option seems faded or grayed out, it means that it has been locked at a higher level. This is most likely for schools that use Zoom for education and businesses that use Zoom for enterprise. At this point, there is nothing you can do except to call your administrator and ask him to enable the feature from his end.

IV. In the advanced cloud recording settings menu, click the "audio transcript" box to select it and then click on "save" to save the recording you created.

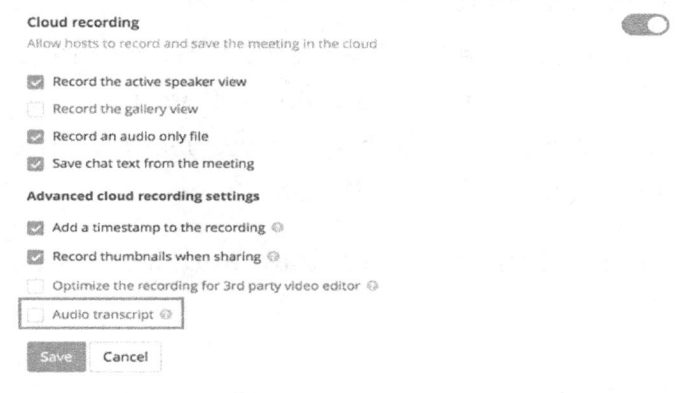

## C. Enabling transcription for a group

I. Sign in to the Zoom platform and from the user window, click on "user management."

II. Click on "group management" and select the desired group name you want to enable transcription. When the group is highlighted, click on the "settings" tab.

III. Seek out the "cloud recording" tab from this menu and make sure that it is enabled. If it is not enabled, toggle the button on and save the change you have effected.

IV. In the advanced cloud recording settings, click on "audio transcript" to enable the feature. Afterward, click "save" to save the changes that you have made to the group.

## Generating a Transcript

Now that you have enabled audio recording and the transcription options available for your paid Zoom account, you must know how to request a transcript of your meetings. To achieve this;

1. Start a meeting or a webinar and start the record feature.

2. Ensure that you have subscribed to any of the paid plans indicated in earlier parts of this book. As stated earlier, the transcript feature is only available to paid users of the platform.

3. Change the settings to make sure that you are recording to the cloud. To achieve this, click on "record to the cloud" after you have clicked on the "record" button. This feature enables the recording you have made to be automatically saved to the cloud storage device once you are done.

4. When you are done with the meeting and you hit the "end" button, you will receive a mail after a while that tells you the recording of your meeting is now available

for access. After a while, you will also receive a notification that the meeting transcript is available for access. Click the notification link to get access to where the transcripts are saved, go through it and make changes where you see fit.

## Ending a Meeting

A lot has been said about setting up your meetings and how you can make sure that you have a seamless experience with your meetings. Now that you have figured out what to do with your meeting and how to run it seamlessly, you are good to go.

So, you are done with your Zoom meeting, but you are not entirely sure how you can end the meeting the right way. Just as the name implies, ending the meeting means bringing a stop, the room's activities, and this automatically ejects every meeting attendee. Here are the steps to follow to bring meetings to a close quickly.

1. Sign in to the Zoom app and set up your meeting as the host.

2. Accept participants and let the meeting go on as planned. When you are done with the meeting, and it is

time to close down the meeting, mouse over to the host controls at the bottom of the Zoom meeting interface and click on "end meeting for all."

3. This option will shut down all activities happening in the room and eject all the meeting room participants.

4. The process is the same, irrespective of your device and the platform you are using. All you need to do is seek out the host controls (which is usually at the bottom of the screen) and click on "end meeting for all" from there. The same action will be carried out for all the meeting participants.

### Scheduling and Using Zoom Webinar

The major difference between a Zoom meeting and a Zoom webinar is the number of people allowed to attend either. While Zoom meetings can take only a few people (a few hundred at most), Zoom webinars can take on up to thousands and even ten thousand people simultaneously (depending on the user's subscription plan). To create a webinar with Zoom, you must be a licensed user of the platform, and you must have gotten the webinar add-on attached to your account. The number of people that can be in your webinar depends on the exact webinar add-on active for you.

To schedule a webinar, this is what you have to do.

1. With your computer, log in to your Zoom portal.

2. From the open window, click on "webinars." This will open you up to a place where you can start setting up your webinar.

3. At the top of the new page that has opened up for you, click on "schedule a webinar.' When you have done this, enter the webinar topic and optionally, a webinar description.

4. Fill in other necessary details, including the date the webinar is to hold, the time it is to hold, the duration it will last for and do not forget to adjust the time zone settings as you do this.

5. If this is going to be a recurrent webinar, click on the "recurrent webinar" option. Note, however, that recurrent webinars can only hold for up to 50 times. When this number is up, you will have to create a new one again. You can set the recurrence frequency to be

daily, weekly, monthly, or to be for no fixed time (in which case you can get to start the webinar at any time you want).

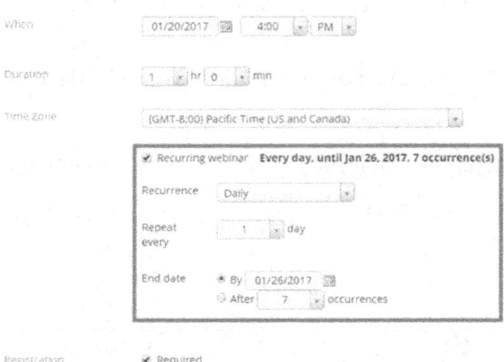

6. If the recurrence of the webinar is daily, weekly, or monthly, go on to specify the end date of the webinars. This is the time when the webinars will stop holding. You can choose a date or the number of times the webinar will hold before it stops.

7. Specify whether or not registration is required for the webinar attendees. If registration is required, you will need to specify when and how they get to register for the webinars; do they register once and can have access to all the webinars, do they need to register for each of the webinars, do they skip the process altogether - in any case, the choice is yours to make and set.

8. Include other webinar settings, then click on the "schedule" button to save the changes you have made to the webinar you just scheduled. You will find a summary of all the webinars you have scheduled on the "manage page for these events."

## Zoom Best Practices

You have gotten all set with the meeting you want to host. However, you must know what you need to do to get the most out of your Zoom meetings. Here are a few practices that can give you the best during Zoom meetings.

1. As a meeting host, never give people your meeting ID (especially if they are not family or trusted friends). This will help you secure yourself from spammers and internet trolls. While you are also at it, remember to password important meetings, enable the waiting room feature, and do all you can to limit confusion and distractions during your meetings.

2. Pay attention to your lighting. This is important for every person attending a Zoom meeting, but it is especially important for you as the meeting host. You

want to use bright enough lights to illuminate your face and still not too harsh on the participants' eyes. As a best practice, make sure that you are facing in the direction of the light. This makes sure that the rays fall on your face, and the reverse is not the case. If you are facing away from the light, it will wrap around you and cast shadows in front of you. This is bad for the visuals of your video conference.

3. See your meetings as to what they truly are; meetings. Although they are remote, please dress the part when you are about to attend a meeting. You will have people looking at you, so you do not want to give off the vibes that you are a tardy person, and you ignore your looks. While you are at it, invest in fixing your surroundings. Since the video component will be on, please ensure that what people will see around you will be visually appealing. It is also one way of ensuring that the perceived value people have of you is on the high side.

4. Test the audio and video components before you start the meeting and accept invitees into the meeting room. Also, take a tour through the meeting room and see to it that everything is on track and that there would not be technical challenges arising during the meeting.

5. While the meeting is on and you are presenting, it may be in everyone's best interest in the meeting room for you to turn off the participants' mics from your end. This way, you do not have to deal with the superimposition of sound from other participants' mics. Also, when people are speaking, put off your mic and encourage others to do the same. For decorum sake, there should only be one person talking at a time.

6. Set up your webcam at eye level or a bit higher. The goal is to make sure that you look comfortable. Keeping the camera down will make it look as though you are struggling to look down as you interact with the people in your meeting room. On the other hand, if the camera is too high up and far away from your face, it does not look too flattering. Try out different positions for the camera and find out what works best for you.

7. Ensure you make eye contact with the people on the other side of the meeting room. One way to look confident as you address people physically is that you make it a point of duty to look them in the eye. As far as virtual meetings and the Zoom app's use is concerned, it does not change. Make eye contact with your meeting attendees by looking at the webcam or the camera at intervals. However, do not overdo this; you do not want to make them feel spooked.

# Chapter 4

## Troubleshooting Most Common Zoom Problems

A lot of things can go wrong with the Zoom platform while you make use of it. The fact that it is an amazing tool does not rule out the possibility that things could easily go haywire, or that you may come across challenges you may be unable to handle.

The essence of this section is to let you in on common problems that you may encounter as you use Zoom and equip you with the basic knowledge to help you fix little bugs and make use of the application without glitches for as much as you can.

Below are the most common problems and their solutions;

**Problem**

Webcam or audio components not working during a meeting.

**Possible cause(s)**

This could result from a lot of issues, including mistakes in meeting setup, technical glitches, and everything else in-between.

**Solution(s)**

Here are a few solutions you need to try out.

1. Before joining a call, make sure that the options that ask you to *turn off my video* or *do not connect to audio are unchecked.* Either of these options will make sure that you do not have video or audio feedback respectively.

2. If your webcam is not showing up, make sure that all other programs that use webcam are all closed down. Zoom is already a demanding tool and having other applications that need the same tools as it does and at the same time, may interfere with getting the best result out of your Zoom meeting experience.

3. To avoid stumbling into problems like these, you may want to conduct a test call ahead of the meeting time. Nothing is more frustrating than your inability to access all the features needed to run your meeting seamlessly. Simply call up someone you trust and feel comfortable with and let the person be your audience

while you double-check that he can hear and see you and your screen from his end. This way, you know if there is any challenge beforehand, and begin to work your way around it.

Check your device's background settings to ensure that the permissions for mic usage and the webcam are not blocked. This is most applicable to you if you are making use of a PC. Go to the device settings and seek out permission options for applications and connected devices to effect these changes if need be.

**Problem**

Meeting lag, or a complete freeze of feedback during meetings.

**Possible cause(s)**

This is usually indicative of a problem with the strength of your internet connection. The weaker your internet connection is, the more likely it is for this to happen to your meeting.

**Solution(s)**

1. Try to move to an area with stronger internet signal reception (if you are using a device that you can move around easily).

2. Test your internet speed before the beginning of the meeting. This way, you know if it is best for you to swap internet service providers or to look for another place to make use of for your meeting. When testing, gun for speeds of 1Mbps download and 800Kbps upload speed (at least). The higher the values you have, the better your meeting experience will be.

3. Improve your video streaming experience by adjusting the quality settings of your video streams. The easiest way to do this is to disable HD options for your video stream. This will, in turn, reduce the amount of data needed for the connection.

**Problem**

Problems with screen sharing

**Possible cause(s)**

This may be due to a few wrong settings, and even internet connection issues can give rise to this glitch during the meeting.

**Solution(s)**

1. Ensure you have a stable internet connection to start with and that you are connected to the call.

2. If you have issues with this during a meeting, you may want to start by turning off the meeting's video component. Doing this will free up some of the bandwidth needed for making sure that you share your screen without issues. Just click on the "stop video" button at the bottom of your screen, then start sharing your screen by clicking the "start screen share" button.

3. If the meeting is not in progress, you may want to try signing out from your account portal and signing back in. Depending on how intense you feel the issue is, you may also want to try uninstalling and reinstalling the application in your device.

Contact support if you have tried out these tips and are still unable to share your screen.

**Problem**

Zoom crashing frequently.

**Possible cause(s)**

This may be a result of several factors, including

- A regional Zoom error in your area.
- Compatibility issues. It could be because the application you downloaded is not compatible with the device you want to use it on.

**Solution(s)**

1. Check the status of Zoom service in your area. You can use Downdetector to see whether there is an issue with the regional state of the service and platform.

2. Try switching up things a little. Make use of the web version of Zoom on your PC. Provided that you have a steady and reliable internet connection, the web version seems to be a bit more reliable and a go-to option when the desktop app begins to malfunction.

3. Make sure that you have installed an application that is compatible with your device. Refer to the compatibility guide as included in the beginning section of this book if you are in doubt or look it up over the internet.

After you have tried the above options and it seems as though the issue is still yet to be solved, you may want to reach out to customer support.

**Problem**

Experiencing challenges with Zoom meeting bombers.

**Possible cause(s)**

This could be because of a few reasons, including

- Insufficient security measures put in place by you, as the meeting host.
- Your meeting ID may have fallen into the wrong hands.

**Solution(s)**

1. Always require a passcode for the meeting attendees. Let this be the first line of security for the meetings you have put in place for the people to attend.

2. Activate the waiting room feature. Although this may be a bit of a hectic task (depending on the number of people you expect to come for your

meeting), it is a security measure that pays off in the long run.

3. Update your version of the application to something more recent. This is because more recent versions have an update to them; end-to-end encryption.

4. As an added precautionary measure, disable participants' microphones, and as the case may be, video feeds too. This is to minimize the number of distractions that can come from their ends and make sure that decorum is maintained within the meeting room.

**Problem**

Popular features missing

**Possible cause(s)**

This could be because you have joined the meeting with a previous version of the application. Constantly, updates are being made to the Zoom app, and with these updates are added features. Most of the features discussed in this book may not be available to users of old versions of Zoom.

**Solution(s)**

1. Update your Zoom version to the most recent version you can find on the internet. Also, make sure that you get something compatible with your device while installing the new app or run the risk of facing the problem (4) above.
2. Log out of the Zoom web host and use a dedicated app for the meeting you are about to host. The desktop app seems to have more features and functionalities than the web version.

**Problem**

Noisy background and sound interference during meetings.

**Possible cause(s)**

This could be because of the superimposition of sound from the mics of other participants during the meeting.

**Solution(s)**

1. Mute all mics from your end as the host. This way, you do not run the risk of experiencing these challenges during your meeting and when the time comes for people to speak, just unmute them.

2. Ask the meeting participants to mute themselves from their end. They need to know how to do this, and as much as it lies within them, they should use quiet places for meetings.

## Chapter 5

### Cool Tips and Tricks to Enhance Productivity With Zoom

Here are a few quick tips and tricks that you need to have up your sleeves to have a smoother experience using the Zoom app.

1. Maximize Zoom keyboard shortcuts. Keyboard shortcuts help you save time on the app. Here are a few common shortcuts you can use on the go.

A. Quick invite; Instead of going through the "invite participants to meeting" process, you can use this shortcut to send a quick invite to the people who need to be in the meeting. On your PC, press Alt+1 to open up the invite window. From this place, you can go through the motions of inviting people to your meeting.

B. Record meetings; Using your PC, type Alt+R to start a recording. When you want to stop/pause the recording you have created, press Alt+P

C. Share screen; Press Alt+Shift+S to start sharing your screen and use Alt+T to pause/resume your sharing.

D. Mute everyone in the call at the same time; Alt+M. This is a great way to save time if you have several people in the meeting room at the same time.

E. To quickly turn off your video feed, use Alt+V.

A number of shortcuts remain. Maximize them to have the best meeting experience and to make sure that you do not spend a lot of time doing what you can achieve with just a few buttons on the keyboard.

2. Unless the video feeds are completely necessary, you will find that it enhances the meeting's audio quality, and the live stream becomes easier for all the meeting participants. So, before the meeting, decide if the video feed is really necessary. If it isn't, do not hesitate to turn it off.

3. If you use Zoom videos for many meetings with people you are not well acquainted with, this is one tip that can help make the name of the person you are talking with always to stay top of mind. Just head over to the meeting's settings, check the video settings and check the box next to *always display participant's name on their videos*.

4. You can make your shortcut usage become even easier by enabling the shortcuts to be used outside of Zoom. This will make the use of those shortcuts that have been discussed to become more powerful, and this applies to you if you usually have other windows open in your device and other applications running while you make use of the Zoom meeting app. This trick allows you to use the same shortcut you use in Zoom when you are in another window on your device.

Go to "settings," click on "keyboard shortcuts" and select "enable global shortcut" to activate this setting for your shortcuts.

5. It can be a bit too much to remember when you have meetings, especially if you tend to have many of them. As a way around this, you can set up a meeting reminder in the Zoom app. This

comes in handy to remind you that you have a meeting slated to hold during a particular time of the day. You can find and activate this feature in the "meeting settings" window of the Zoom app. Toggle the button on to make sure that you do not miss out on meetings.

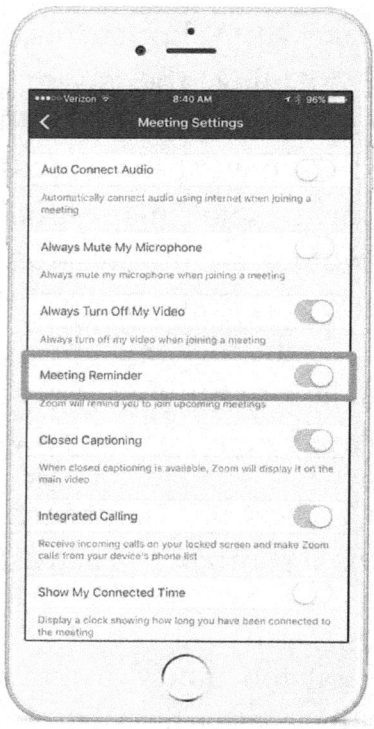

6. You can touch up your appearance in the application. This is very handy to ensure that you do not look haggard or haphazard as you attend meetings, especially if you need to make a good appearance in the meeting. To achieve this, go to "video settings" and check the box next to the "touch up my appearance" option available in the menu for you.

7. Zoom works seamlessly with many other applications and tools. When integrated, these tools can give a powerful effect that can take your meetings to the next level.

Some of these tools are;

A. Zapier; is a great scheduling tool that you can connect to your Zoom. It is also great for reminding participants that a meeting is set to hold at a specified time.

B. Slack; you can start meetings right from here and use it as a collaborative tool for your teams.

8. With updated versions of Zoom, you can *change your background*. This is a very handy feature if

you are looking to make a great impression and you know that your surroundings will not quite cut it. This is an easy hack, and to get it done, all you need to do is go to "settings," find and click on the "virtual background" tab. From there, you can select the picture/upload the picture you want to use for your meetings.

9. Turning on the gallery view allows you to see everyone in the meeting room at the same time. Instead of just seeing the person that is speaking per time, with this option, you can keep an eye on all the people in the meeting room at the same time. To activate this, click on the "gallery view" tab at the top of the meeting window. Note, however, that this is only available for a meeting with 49 attendees or lesser.

# The end... almost!

Hey! We've made it to the final chapter of this book, and I hope you've enjoyed it so far.

If you have not done so yet, I would be incredibly thankful if you could take just a minute to leave a quick review on Amazon

Reviews are not easy to come by, and as an independent author with a little marketing budget, I rely on you, my readers, to leave a short review on Amazon.

Even if it is just a sentence or two!

So if you really enjoyed this book, please...

>> Click here to leave a brief review on Amazon.

I truly appreciate your effort to leave your review, as it truly makes a huge difference.

# Chapter 6

## Zoom Frequently Asked Questions (FAQs)

Here are few FAQs that people have asked over time, and answers to these questions;

**Question**

Do I need an account to use Zoom?

**Answer**

You do not necessarily need an account if all you are doing is to join a meeting as a participant. However, if the host of the meeting has restricted access to the meeting by enabling the use of authentication profiles, you will need to have an account and sign in to the account to access the meeting.

On the other hand, you need to have a Zoom account if you are going to be a host of meetings. Creating an account is a simple task. Just follow the instructions in the beginning part of this book to set up your own account immediately

**Question**

How do I sign up for the Zoom service?

**Answer**

This is simple. All you need to do is log on to Zoom.us/signup and follow the easy steps that are outlined on the portal.

**Question**

How do I activate my device's audio during a meeting so that I can hear what the speaker is saying?

**Answer**

On most devices, all you need to do is to hit the "join audio" button. Many times, you will see this button at the bottom of the screen. A few other variants include join with computer audio, audio, etc (depending on the device you are dialing in with).

**Question**

Can I automatically add users to my account?

**Answer**

Yes, you can. Business, Education, and Enterprise accounts can automatically add a user by adding an associated domain to their account. The associated domain feature is a premium feature that uses an organization's email address domain name (@Zoom.us) to add or auto-create users whose emails are attached to that account (for example, Julia@Zoom.us).

**Question**

Can I add other people to help me manage my account?

**Answer**

Yes, you can. A registered user of the application can add administrators or users who have special roles to their account to help them run specific tasks on the platform.

**Question**

Must I have a webcam to join in a meeting?

**Answer**

No. You must not have a webcam to be in a meeting. However, if you will share video feedback at some point, you must have a webcam to be a part of the meeting. If you are sure that you won't be doing this, you may want to just dial in via the meeting's audio component.

**Question**

How do I hold a webinar on Zoom?

**Answer**

To do this, you first need to purchase a webinar license. Get started by logging on to Zoom.us/billing to find all the licenses that are available for purchase.

After purchasing a license, log on to the webinar scheduling page here at Zoom.us/webinar/list. Create and schedule your webinar by following the steps that have been discussed in the earlier parts of the book.

**Question**

Can I reset my Zoom password after I have forgotten it?

**Answer**

Yes, you can. To get started, log on to the password reset page at Zoom.us/forgot password

**Question**

What is the difference between a basic and a licensed user of the service?

**Answer**

A basic user is the user of Zoom's free package. As a basic user, you are entitled to holding a meeting with up to 100 people at the same time. If up to three people show up for the meeting, as a basic user, the meeting will time out after 40 minutes and you would have to end the meeting or create a new meeting room for people to dial in to. Basic users do not have access to many of the advanced features available for the platform's premium users.

On the other hand, a licensed user is a paid user that can host meetings beyond the 40 minutes time limit. A licensed user is entitled to host meetings with up to 100 people by default. They can have access to additional people when they have purchased additional licenses for their meeting capacity. They have access to many more features and can make the most out of the platform instead of the free user who only has access to the platform's basic features.

**Question**

Can I use the Google SSO feature with Zoom?

**Answer**

Yes. You can use Google's Single Sign On feature with
your Zoom account.

# Conclusion

The Zoom application is a go-to tool for you if you are looking to leverage the internet's power to host virtual meetings and not be restricted by location and place. Coupled with that, it comes with many tools and features that can help you make the most out of every meeting.

This book has taken you through a tour of the application and has shown you how you can make the most out of your Zoom account and the meetings you plan to hold. From setting up your account to getting started with your first meeting until you become a skilled user of the platform - almost everything you will need to make the most out of the platform has been discussed in great detail here.

As you set up for your next meeting, refer to the content of this book. They will make the journey easier for you, and see to it that you and the participants in your meeting room have a great meeting experience.

CPSIA information can be obtained
at www.ICGtesting.com
Printed in the USA
BVHW071824191020
591325BV00004B/1381

9 781952 597381